Barndominium Floor Plans and Designs: Build Your Dream Rural Residence - A Comprehensive Collection of Practical Floor Plans & 3D Renderings

Tony Bobin

Scan the QR Code and access your 3 bonuses in digital format

Bonus 1: Basic Furnishing Guide for Your Barndominium

Bonus 2: Annual Maintenance Tips for Your Barndominium

Bonus 3: Utility Setup Guide for Your Barndominium

Table of Contents

Introduction.. 7

Tranquility Terrace: A Modern Mono Cabin Blueprint 9

The Neru Cabin.. 17

Introduction to the Hygge Cabin: Embrace the Essence of Nordic Tranquility.. 25

Zen House – Elegance Defined 33

Chi Cabin: A Modern Sanctuary 41

The Tranquil Nook Cabin.. 48

The Homestead Sanctuary 59

The Comforts of Contemporary Living 67

The Urban Nest 77

Conclusion... 87

Introduction

Welcome to the threshold of possibility where your dream of rural living meets the tangible reality of building your own barndominium. "Barndominium Building Mastery" is not just a guide; it is your comprehensive blueprint for transforming a vision into a sturdy, stylish, and sustainable home. This book is crafted for dreamers, planners, and doers—anyone who sees the beauty in sprawling landscapes and the charm in custom-built living spaces.

Why a barndominium? These structures combine the rugged functionality of a barn with the refined comfort of a condominium, offering a unique architectural blend that supports a variety of lifestyles, from the rustic to the remarkably modern. Whether you're drawn to barndominiums for their aesthetic appeal, their cost-effectiveness, or their adaptability, this guide will walk you through every step necessary to create a space that is uniquely yours.

Within these pages, you will find detailed, step-by-step instructions that demystify the complexities of building from the ground up. From selecting the perfect plot of land to laying the last tile, every phase of construction is covered. You'll gain insights into managing costs without compromising on quality, designing with foresight, and navigating the myriad regulations that come with construction.

Moreover, this book is imbued with strategies to enhance your project's success, including exclusive tips on customization options that cater to diverse needs and preferences. With five detailed residence floor plans, complete with renderings, and additional bonuses like budgeting strategies and a myth-busting section, this guide is designed to be both practical and inspiring.

As you embark on this exciting journey of building your barndominium, remember that every chapter in this book is a stepping stone towards creating a home that celebrates your individuality while embracing functionality. Get ready to lay the foundations of a home that's built to reflect your dreams and to stand the test of time.

Let's begin this extraordinary adventure together, transforming blueprints into breathtaking realities. Welcome to "Barndominium Building Mastery"—where your dream home awaits.

Tranquility Terrace: A Modern Mono Cabin Blueprint

Welcome to "Tranquility Terrace," a blueprint that encapsulates more than just the architectural nuances of building a modern Mono Cabin—it is a gateway to a lifestyle of serene, simplified elegance. Nestled within these pages, you will find not only plans but also an invitation to reimagine what your living space can be. Designed with precision by the innovative minds at UtomoStudio, this 450-square-foot cabin emerges as a masterclass in the art of compact luxury and minimalist design, tailored perfectly for those looking to embrace a quieter, more intentional way of living.

As you unfold the pages of this project, each section of the cabin is introduced with detailed renderings that guide you from the welcoming threshold through a seamlessly integrated living space, and into a tranquil bedroom sanctuary. This journey is carefully curated to show how each area can be maximized for both function and comfort within a modest footprint.

The architectural story of the Mono Cabin begins with its striking entrance—large, inviting double French doors that not only enrich the facade but also blur the lines between the lush outdoors and the cozy indoors. This design choice is aimed at expanding the visual and physical living space, extending it out onto a verdant terrace, where the boundaries of home feel limitless.

The aesthetic of the cabin speaks of understated elegance, featuring pristine white plastered walls that offer a canvas of calm against the rustic charm of classic Spanish terracotta roof tiles. This blend of traditional materials and modern design principles is highlighted further by the distinctive silhouette of the cabin's gable roof. A functional yet stylish parapet wall encases the structure, providing both privacy and a visual anchor that draws the eye upward, enhancing the structure's vertical dimension.

Inside, the living area is conceived as the heart of the home. It is an open, fluid space where the kitchen and dining areas coalesce with the relaxation zone, creating an environment that encourages interaction and engagement. Central to this area is a robust, elegantly designed fireplace that stands as much as a source of warmth as it is a centerpiece, around which daily life unfolds.

Above, the pitched ceiling arches gracefully, expanding the sense of space vertically and inviting natural light to permeate the cabin, creating an interior that feels both expansive and intimate. The design meticulously integrates elements like built-in shelves and concealed storage, showcasing how functionality can be woven seamlessly into the fabric of stylish, modern living.

"Tranquility Terrace" offers more than just a living space—it presents a philosophy of life. It invites you to construct a home that reflects a commitment to beauty, simplicity, and functionality. As you delve into this project, consider it not just as a construction endeavor, but as a craft—a process of creating a sanctuary that embodies peace, comfort, and modern rustic charm. Embrace this journey of transformation with "Tranquility Terrace," and let it guide you toward building a space where every square foot is infused with purpose and every room opens the door to a more mindful, fulfilling existence.

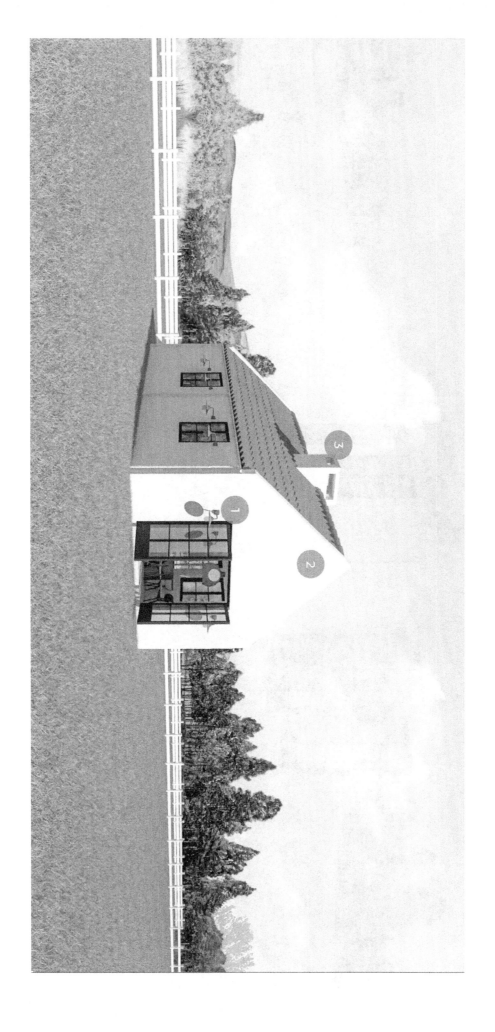

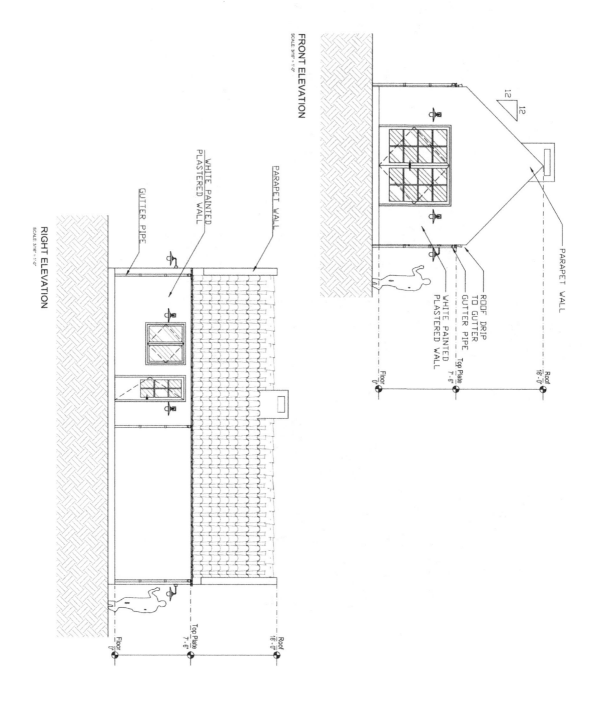

FRONT ELEVATION
SCALE: 3/16" = 1'-0"

RIGHT ELEVATION
SCALE: 3/16" = 1'-0"

GUTTER PIPE

WHITE PAINTED
PLASTERED WALL

PARAPET WALL

PARAPET WALL

WHITE PAINTED
PLASTERED WALL

ROOF DRIP
TO GUTTER
GUTTER PIPE

Floor
0'

Top Plate
7'-6"

Roof
16'-0"

Floor
0'

Top Plate
7'-6"

Roof
16'-0"

14

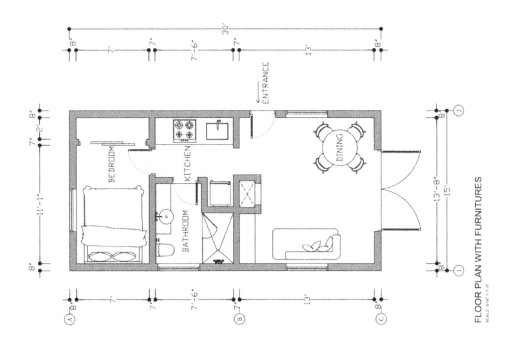

FLOOR PLAN WITH FURNITURES
SCALE 3/16"=1'-0"

ENTRANCE

BEDROOM

KITCHEN

BATHROOM

DINING

30'

8" 7' 7" 7'-6" 7" 13' 8"

11'-1" 7'-2'-8" 8'

A B C

13'-8" 15' 8" 8"

1 2

The Neru Cabin

Welcome to the Neru Cabin, where contemporary design meets the rustic charm of rural living. This blueprint isn't just a layout; it's a vision of life harmoniously integrated with nature, designed to foster a connection with the outdoors while providing all the comforts of modern living. As you step through the large sliding glass doors, you're invited into a space where boundaries between the inside and the expansive world beyond blur into one seamless experience.

The exterior of the Neru Cabin is a blend of sophistication and simplicity. Vertical white-painted wooden sidings give the structure a clean, streamlined look, while black metal roofing stretches down to meet the walls, offering a striking contrast that complements the natural surroundings. This choice of materials is not only aesthetically pleasing but also ensures durability against diverse weather conditions. The asymmetrical gable roof not only adds to the visual interest but also enhances the interior's spatial dynamics, creating an airy, open feel that belies the cabin's modest footprint.

Inside, the cabin's layout is thoughtfully designed to maximize space and functionality. The living area, though compact, feels spacious due to smart furniture choices like a low sofa which helps maintain an uncluttered line of sight across the room. The dining area features a large wooden table that comfortably seats four, making it an ideal setting for family meals or small gatherings.

Natural lighting is a pivotal element of the Neru Cabin's design. Skylights strategically placed along the high sloped ceiling illuminate the space with daylight, reducing the need for artificial lighting and offering a constant connection to the sky above. This not only helps in energy conservation but also enhances the overall well-being of the inhabitants by aligning indoor living with the natural rhythms of the day.

Sustainability is at the core of the Neru Cabin's design philosophy. The materials, from the paint to the roofing, are selected with environmental impact in mind, ensuring that the cabin not only blends into its setting but also contributes to the conservation of resources. The design encourages a lifestyle that values simplicity and efficiency, principles that are becoming increasingly important in our fast-paced world.

This project guide will take you through every aspect of building the Neru Cabin, from laying the foundation to the finishing touches of the interior. It provides detailed instructions and expert tips to ensure that every phase of construction goes smoothly, whether you're a seasoned builder or a first-time DIYer. With this comprehensive guide, building your dream rural mansion becomes not just a possibility, but a tangible project ready for realization.

Embrace the blend of modern design and rustic charm with the Neru Cabin, your gateway to a lifestyle that celebrates the beauty of rural living, the joy of simplicity, and the satisfaction of building a space that is truly your own.

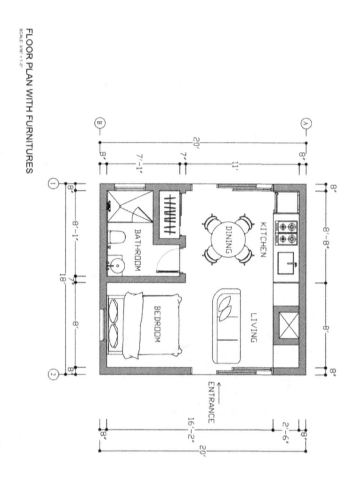

FLOOR PLAN WITH FURNITURES
SCALE 3/16" = 1'-0"

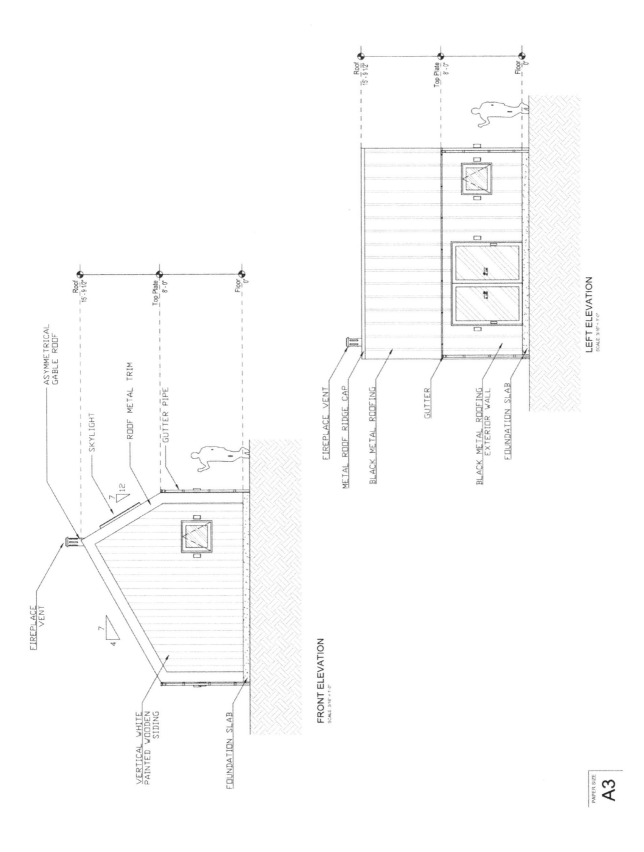

ASYMMETRICAL
GABLE ROOF

SKYLIGHT

ROOF METAL TRIM

GUTTER PIPE

FIREPLACE
VENT

VERTICAL WHITE
PAINTED WOODEN
SIDING

FOUNDATION SLAB

Roof
15'-9 1/2"

Top Plate
8'-0"

Floor
0"

12
7

7
4

FRONT ELEVATION
SCALE 3/16" = 1'-0"

FIREPLACE VENT

METAL ROOF RIDGE CAP

BLACK METAL ROOFING

GUTTER

BLACK METAL ROOFING
EXTERIOR WALL

FOUNDATION SLAB

Roof
15'-9 1/2"

Top Plate
8'-0"

Floor
0"

LEFT ELEVATION
SCALE 3/16" = 1'-0"

PAPER SIZE
A3

23

Introduction to the Hygge Cabin: Embrace the Essence of Nordic Tranquility

The **Hygge Cabin** embodies the Scandinavian ethos of comfort, functionality, and connection with nature, offering a blueprint for those looking to infuse their lives with a serene, content quality through their living environments. With its modest 540 square feet, the **Hygge Haven** is more than just a physical structure; it's a manifestation of a lifestyle philosophy deeply rooted in simplicity and tranquility.

This project introduces you to a carefully designed floor plan that balances aesthetic and functional elements to promote a relaxed lifestyle. The cabin features a spacious bedroom that seamlessly extends into a cozy balcony, creating an inviting space for quiet mornings or intimate evenings under the stars. This thoughtful design encourages residents to slow down and connect with the outdoors, enhancing their living experience.

The core of this home is its open-plan living area, where large sliding glass doors not only bring in an abundance of natural light but also provide panoramic views of the surrounding landscape. The strategic placement of a low-profile bed and a floating wooden desk maximizes the sense of space, making the compact area feel both expansive and intimate. A well-equipped kitchen and dining area allow for the preparation and enjoyment of home-cooked meals, elevating daily dining into a delightful experience.

Aesthetic considerations such as vertical wooden siding and sleek black metal roofing reflect a contemporary style while honoring traditional craftsmanship, ensuring durability and timeless appeal. The inclusion of skylights and strategically placed square windows not only enhances the cabin's energy efficiency but also offers beautifully framed views of the dynamic skies above.

Each detail of the **Hygge Haven** has been meticulously curated to ensure that it does not merely function as a house, but as a true home—a retreat from the hectic pace of modern life, offering a return to what truly enriches us: simplicity, quality, and comfort. This project transcends the conventional blueprint to offer a pathway to a richer, more deliberate lifestyle. It invites you to embrace the slower pace, breathe in the calm, and live deeply the hygge way of life, making every moment at home a testament to well-being and happiness.

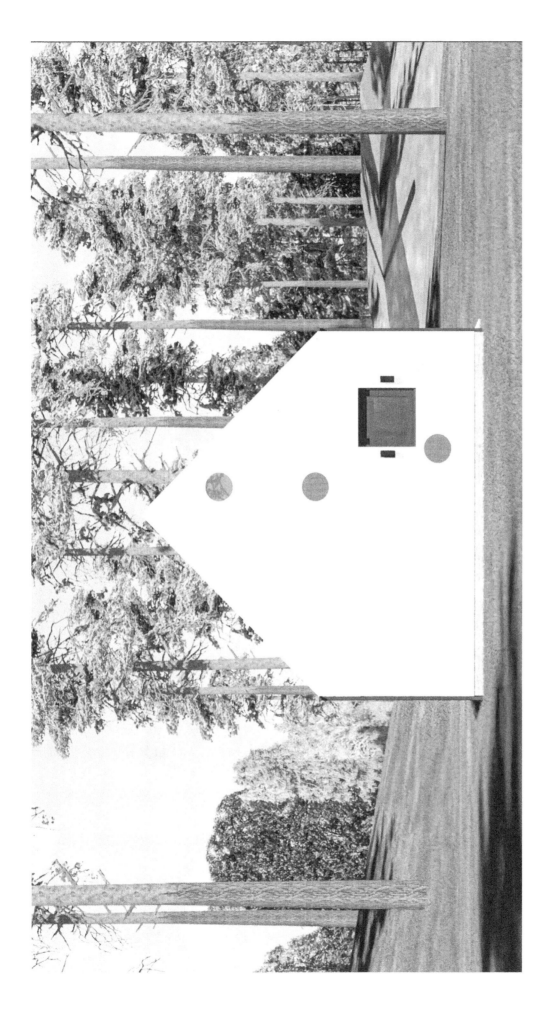

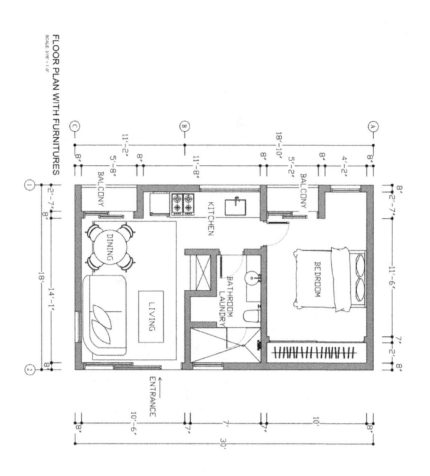

FLOOR PLAN WITH FURNITURES
SCALE 3/16" = 1'-0"

BALCONY

KITCHEN

BALCONY

DINING

LIVING

BATHROOM LAUNDRY

BEDROOM

ENTRANCE

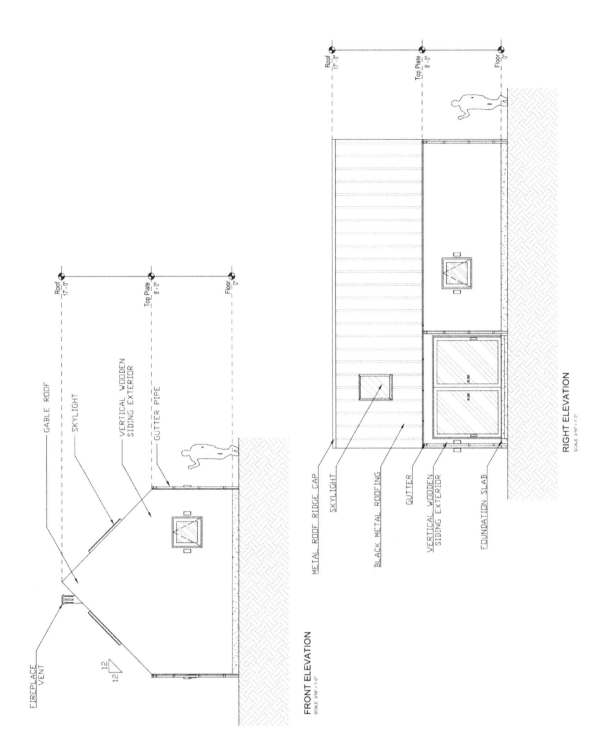

GABLE ROOF

SKYLIGHT

VERTICAL WOODEN SIDING EXTERIOR

GUTTER PIPE

FIREPLACE VENT

12
12

FRONT ELEVATION
SCALE 3/16" = 1'-0"

Roof
17'-0"

Top Plate
8'-0"

Floor
0"

METAL ROOF RIDGE CAP

SKYLIGHT

BLACK METAL ROOFING

GUTTER

VERTICAL WOODEN SIDING EXTERIOR

FOUNDATION SLAB

RIGHT ELEVATION
SCALE 3/16" = 1'-0"

Zen House – Elegance Defined

Welcome to the Zen House, a design that elegantly captures the essence of modern minimalism while offering a tranquil retreat in a compact 579-square-foot layout. This project, designed for those who appreciate the beauty of simplicity, features a striking combination of sharp, clean lines and a monochrome color palette that complements its serene setting. The exterior, characterized by its sleek black metal roofing and pristine white walls, creates a visually striking contrast that enhances the structure's minimalist aesthetic. Large windows are strategically placed to not only flood the interior with natural light but also to frame the picturesque views of the surrounding landscape, blurring the lines between indoor and outdoor living.

Inside, the Zen House maximizes every inch of its space without compromising on style or functionality. The open-plan layout seamlessly integrates the living, dining, and kitchen areas, making it an ideal space for both relaxation and social gatherings. The kitchen is equipped with state-of-the-art appliances nestled under a dropped ceiling, which adds a distinct character to the area while also defining the cooking space. Adjacent to it, the dining area, though compact, comfortably accommodates four, offering a perfect setting for intimate dinners or morning reflections.

The interior design emphasizes sustainability and efficiency. From the flooring to the ceiling, each material is chosen not only for its aesthetic appeal but also for its low environmental impact and durability. This commitment to sustainability is mirrored in the house's energy-efficient fixtures and fittings, which ensure that the Zen House is as cost-effective as it is beautiful.

Customization is at the heart of the Zen House design, allowing prospective homeowners to infuse their personal style into the finishes and details. Whether it's selecting bespoke cabinetry for the kitchen or choosing eco-friendly paint for the interiors, every element can be tailored to meet the owner's preferences, ensuring that each Zen House reflects the unique tastes of its inhabitants.

This project isn't just about building a house; it's about creating a home that stands as a sanctuary of peace and creativity. It's an invitation to embrace a minimalist lifestyle without sacrificing luxury and comfort. The Zen House is designed to be more than a dwelling; it is a space where life slows down, and every moment is savored. For those looking to build a personalized retreat that embodies modern elegance and tranquil living, the Zen House provides a perfect blueprint to start with. Dive into this project, and begin the journey of constructing not just any home, but your dream rural mansion.

FLOOR PLAN WITH FURNITURES
SCALE 3/16" = 1'-0"

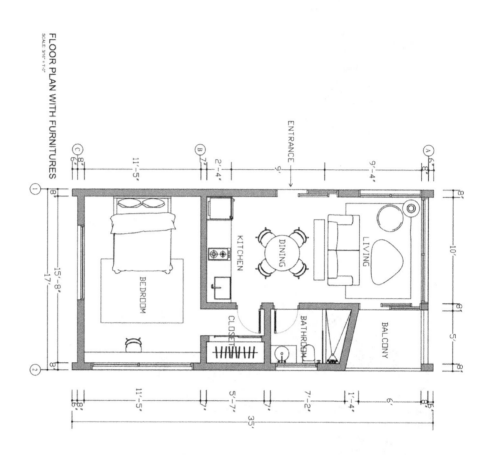

ENTRANCE

KITCHEN
DINING
LIVING
BEDROOM
CLOSET
BATHROOM
BALCONY

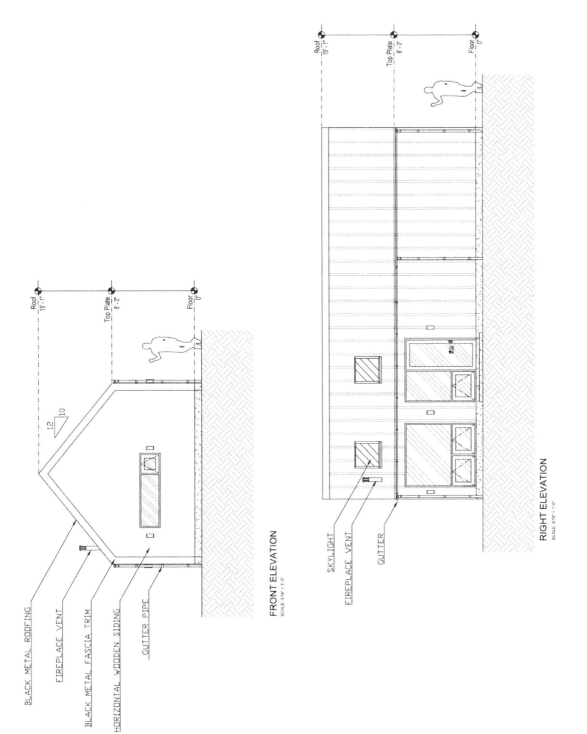

BLACK METAL ROOFING

FIREPLACE VENT

BLACK METAL FASCIA TRIM

HORIZONTAL WOODEN SIDING

GUTTER PIPE

FRONT ELEVATION
SCALE: 3/16" = 1'-0"

Roof
15'-1"

Top Plate
8'-0"

Floor
0'

12
10

SKYLIGHT

FIREPLACE VENT

GUTTER

RIGHT ELEVATION
SCALE: 3/16" = 1'-0"

Roof
15'-1"

Top Plate
8'-0"

Floor
0'

39

Chi Cabin: A Modern Sanctuary

Welcome to the "Chi Cabin," a sanctuary crafted not just as a home but as a retreat into nature's embrace, designed for those who yearn for a minimalist lifestyle without sacrificing the nuanced comforts of modern living. This distinctive barndominium stands as a testament to sleek, contemporary design married perfectly with functional living spaces, making it an impeccable choice for families or individuals who cherish the balance of aesthetic simplicity and practical utility.

At the heart of the Chi Cabin's design philosophy is an open-plan layout that not only maximizes space but also fosters a seamless interaction between the kitchen, dining area, and living spaces. This thoughtful configuration ensures that whether you're preparing a meal or lounging with a book, you're never too far from the beautiful vistas outside your window or the company of loved ones. The expansive use of glass in this design not only invites an abundance of natural light but also frames the captivating panoramic views of the surrounding landscape, effectively blurring the lines between indoor comfort and outdoor serenity.

Built with sustainability at its core, the Chi Cabin utilizes eco-friendly materials and incorporates energy-efficient solutions to minimize its environmental footprint while maximizing comfort. The interior atmosphere is designed to be both cozy and airy, promoting a sense of tranquility and inspiration through its spacious yet intimate settings.

The Chi Cabin is more than just a house; it's a lifestyle choice. It caters to those who love to entertain, with ample space for gatherings, yet it also provides intimate nooks for those seeking solitude in the quiet whispers of the natural world. It represents a beacon of modern architectural ingenuity, a place where every element is designed to enhance the quality of daily living while offering a peaceful retreat from the bustling world outside its walls.

Whether you find joy in hosting vibrant social gatherings or in the quiet contemplation of scenic views, the Chi Cabin is designed to accommodate and inspire. It stands as a bold statement in modern living, where design meets function, and style meets sustainability.

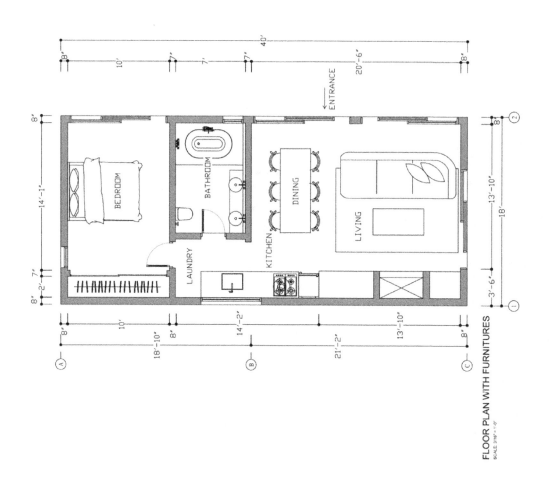

FLOOR PLAN WITH FURNITURES

SCALE 3/16" : 1'-0"

BEDROOM

BATHROOM

LAUNDRY

KITCHEN

DINING

LIVING

ENTRANCE

46

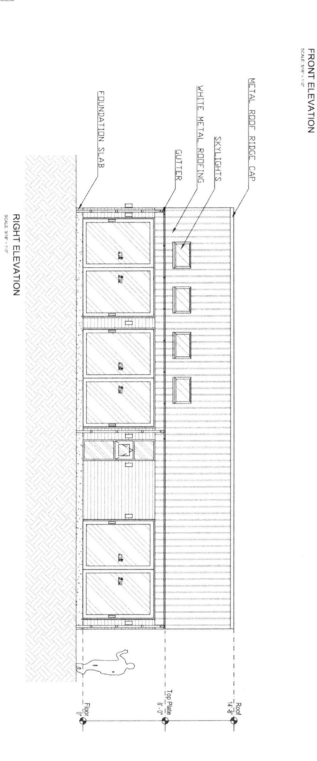

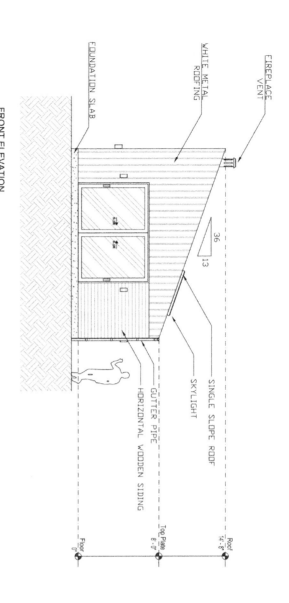

FRONT ELEVATION
SCALE 3/16"=1'-0"

FOUNDATION SLAB

WHITE METAL ROOFING

FIREPLACE VENT

36

13

SINGLE SLOPE ROOF

SKYLIGHT

GUTTER PIPE

HORIZONTAL WOODEN SIDING

Floor 0"

Top Plate 8'-0"

Roof 14'-8"

RIGHT ELEVATION
SCALE 3/16"=1'-0"

FOUNDATION SLAB

WHITE METAL ROOFING

SKYLIGHTS

METAL ROOF RIDGE CAP

GUTTER

Floor 0"

Top Plate 8'-0"

Roof 14'-8"

Scan the QR Code and access your 3 bonuses in digital format

Bonus 1: Basic Furnishing Guide for Your Barndominium

Bonus 2: Annual Maintenance Tips for Your Barndominium

Bonus 3: Utility Setup Guide for Your Barndominium

The Tranquil Nook Cabin

Welcome to "The Tranquil Nook Cabin," a serene retreat meticulously designed to epitomize the art of minimalist living without sacrificing comfort or aesthetic pleasure. Nestled within a cozy corner of your imagination, this blueprint is a gateway to a quieter, more reflective way of life, perfect for those who cherish the simplicity and beauty of rural living.

This design is a masterclass in utilizing space efficiently, ensuring that its compact structure never feels cramped but rather surprisingly spacious. The architectural plan cleverly uses vertical layouts to heighten the sense of openness, with lofty ceilings and expansive windows that not only enrich the living space with abundant natural light but also offer panoramic views of the surrounding landscape. Each room flows seamlessly into the next, with a layout that encourages tranquility and ease of movement.

Moreover, "The Tranquil Nook Cabin" integrates eco-friendly materials and energy-efficient solutions, reflecting a commitment to sustainability that resonates deeply in today's environmentally conscious world. From the timber frames to the energy-saving fixtures, every aspect of the cabin is designed with the environment in mind, ensuring that your home not only adds to your quality of life but also contributes positively to the ecosystem.

This project is more than just a building; it's a promise of a peaceful life, offering a sanctuary from the fast pace of modern living. It's a place where every sunset and sunrise stretches time, allowing you to savor each moment more deeply. With "The Tranquil Nook Cabin," you step into a world where each day ends with gratitude and every morning starts with the beauty of the natural world right at your doorstep. It's an invitation to slow down, breathe deeply, and enjoy life in its purest form.

Embrace the charm of living in a space that blends seamlessly with nature, where the boundaries between indoors and outdoors blur, creating a living experience that is both exhilarating and calming. "The Tranquil Nook Cabin" isn't just a residence; it's a lifestyle, a philosophy of living that respects the rhythms of nature and the depth of human serenity. This cabin is an architectural melody composed in the key of simplicity and sustainability, offering a refuge that nourishes the soul and rejuvenates the spirit.

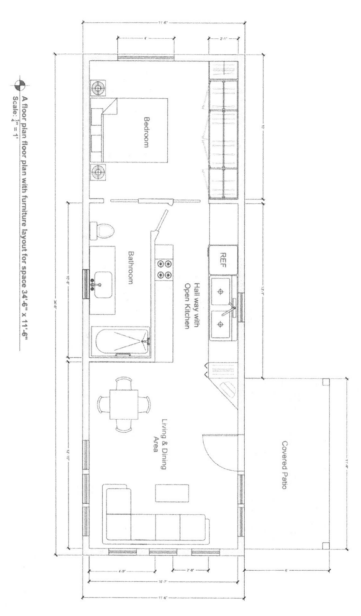

A floor plan floor plan with furniture layout for space 34'-6" x 11'-6"

Scale: $\frac{1}{4}$" = 1'

Bedroom

Bathroom

Hall way with
Open Kitchen

REF

Living & Dining
Area

Covered Patio

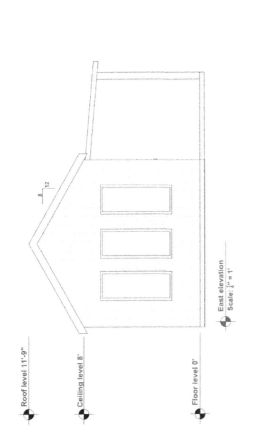

Roof level 11'-9"

Ceiling level 8'

Floor level 0'

8 | 12

East elevation
Scale: ¼" = 1'

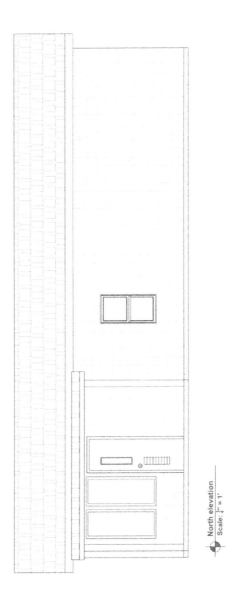

North elevation
Scale: ¼" = 1'

57

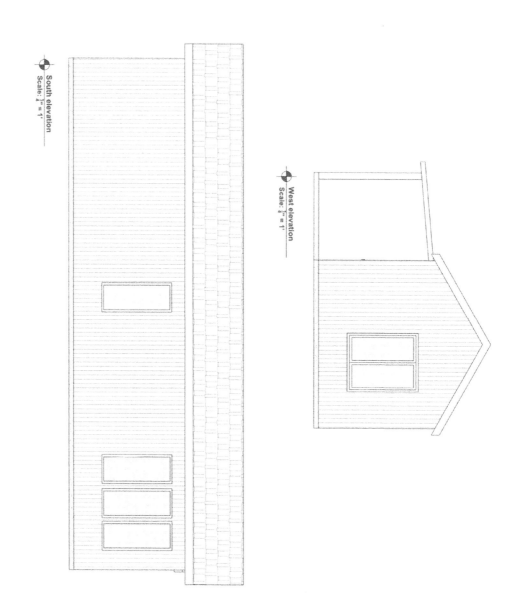

South elevation
Scale: $\frac{1}{4}$" = 1'

West elevation
Scale: $\frac{1}{4}$" = 1'

The Homestead Sanctuary

Welcome to the blueprint of what could be your next dream home—a 2,000 sqft barndominium that's not just a residence but a sanctuary designed to provide the perfect balance between modern living and rustic charm. This three-bedroom house combines the essence of rural tranquility with the efficiencies of contemporary design, making it an ideal choice for those looking to craft a life away from the hustle of city noise.

At the heart of this design is a spacious layout that supports both privacy and communal activities. The open-plan living area, which seamlessly integrates the kitchen and dining spaces, serves as the epicenter of family life, encouraging interaction and shared experiences. Each bedroom is thoughtfully positioned to offer retreat and restorative space, promoting a sense of peace and personal space essential for any family or individual.

Designed by a conscientious architect, every inch of this barndominium has been meticulously planned to enhance livability while respecting the natural environment it's set to inhabit. From the positioning of windows to maximize light and air circulation to the inclusion of a functional yet stylish mechanical/gas room, no detail has been overlooked. The strategic placement of closets and storage solutions ensures that the aesthetic is clean and uncluttered, which is vital in a home that champions serene living.

The exterior views included in the project offer a visual understanding of the house's interaction with its surroundings, illustrating a design that's as beautiful from the outside as it is functional on the inside. Whether it's the inviting front patio or the generous back deck, outdoor living spaces are treated as extensions of the indoor, a testament to a design philosophy that values the outdoors as much as the interior.

For those envisioning a life that combines the simplicity of rural living with the necessities of modern comfort, this barndominium is more than a house—it's a potential home that promises a lifestyle of tranquility, convenience, and aesthetic pleasure. As you turn each page of these blueprints, consider not just the walls and floors that frame this space, but the laughter, memories, and quiet moments that could fill it. Here, in the Homestead Sanctuary, you're not just building a house; you're crafting a home.

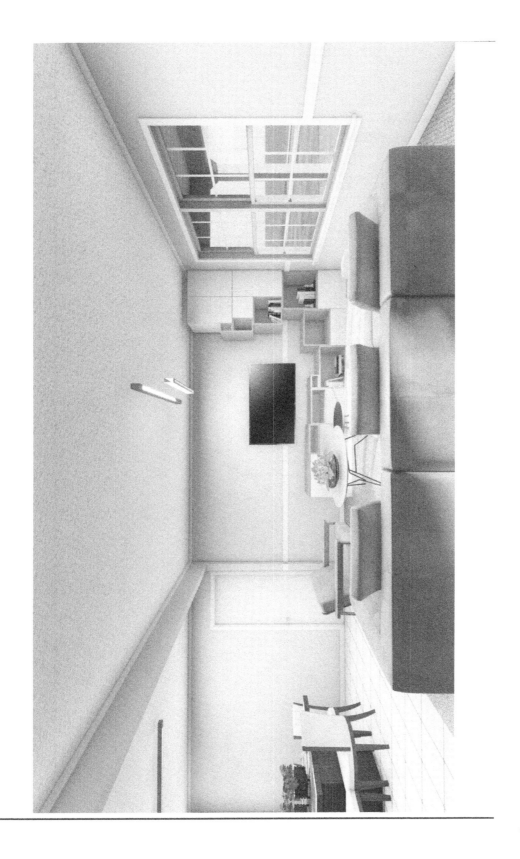

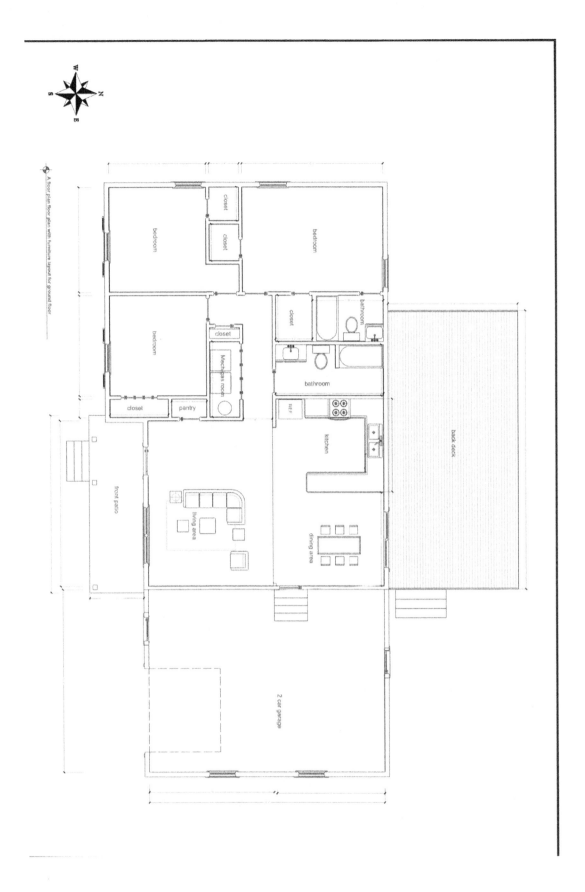

A floor plan floor plan with furniture layout for ground floor

bedroom

closet

closet

bedroom

bedroom

closet

bathroom

closet

bathroom

closet

Mech/gas room

closet

pantry

kitchen

REF

back deck

front patio

living area

dining area

2 car garage

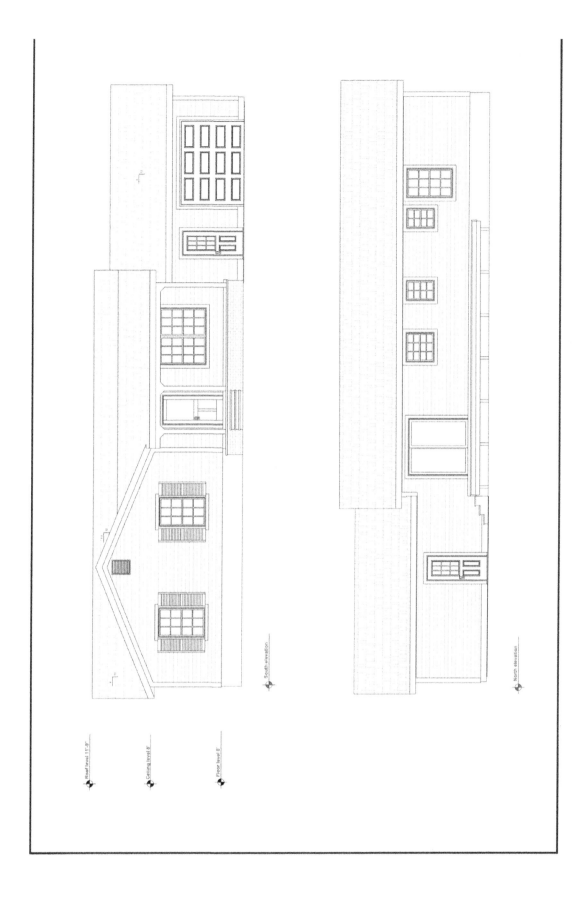

South elevation

North elevation

Roof level 11'-3"

Ceiling level 8'

Floor level 0'

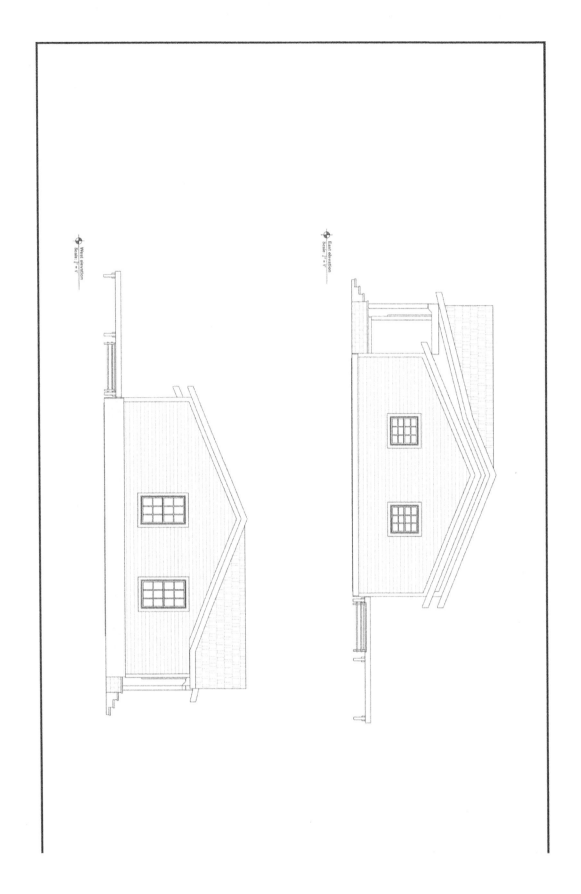

West elevation
Scale 1" = 1'

East elevation
Scale 1" = 1'

The Comforts of Contemporary Living

Embark on a journey to redefine rural elegance with this meticulously designed barndominium, crafted for those who yearn for a seamless blend of modernity and pastoral charm. Envision yourself in a space where every inch serves a purpose, from the cleverly allocated storage areas to the expansive living quarters that promote a lifestyle of comfort and convenience.

This two-bedroom barndominium is the epitome of thoughtful design, offering ample space for both family living and entertaining guests. Each bedroom is strategically placed to ensure privacy while providing easy access to the home's communal areas, fostering a harmonious balance between solitude and social interaction. The inclusion of multiple closets and a dedicated mechanical/gas room highlights a practical approach to rural living, ensuring that functionality and aesthetics coexist beautifully.

At the heart of this home lies a spacious kitchen, dining, and living area, designed not just for daily activities but as a central hub for making memories. The open floor plan encourages a free flow of conversation and activity, making it perfect for gatherings or quiet evenings at home. The detailed floor plans not only show the layout but also suggest furniture arrangements, offering potential homeowners guidance on how to best utilize the space.

The exterior elevations reveal a structure that is both majestic and inviting, with a façade that mirrors the innovation and style of the interior. From the southern charm of the front patio to the practical elegance of the east and west elevations, every angle of this barndominium has been crafted with precision and care.

This project is more than just a set of blueprints; it is a guide to building a home that balances the rustic appeal of rural living with the demands of modern life. It promises a dwelling that is both a sanctuary and a statement, a place where every detail has been tuned to the rhythms of country living but adapted to the modern age. This is where contemporary design meets rural tranquility, creating a living space that is as beautiful as it is functional, as inviting as it is exclusive.

From the materials used in the facade to the layout of the living spaces, every element has been chosen with both sustainability and style in mind. The plans include options for energy-efficient fixtures and renewable energy solutions, ensuring that the home is not only beautiful but also environmentally conscious. The potential for customization in finishes and fittings allows each homeowner to infuse their personal style into the final design, making each build a unique reflection of its owner.

This project invites you to imagine a life where beauty and functionality are perfectly aligned, where each day is lived in spaces designed with care and precision. For those looking to build a home that reflects their values and lifestyle, this project is not just a plan, but a pathway to achieving a dream rural mansion. It's an invitation to build a future where home is not just a place, but a retreat that reflects the best of both rural charm and modern convenience.

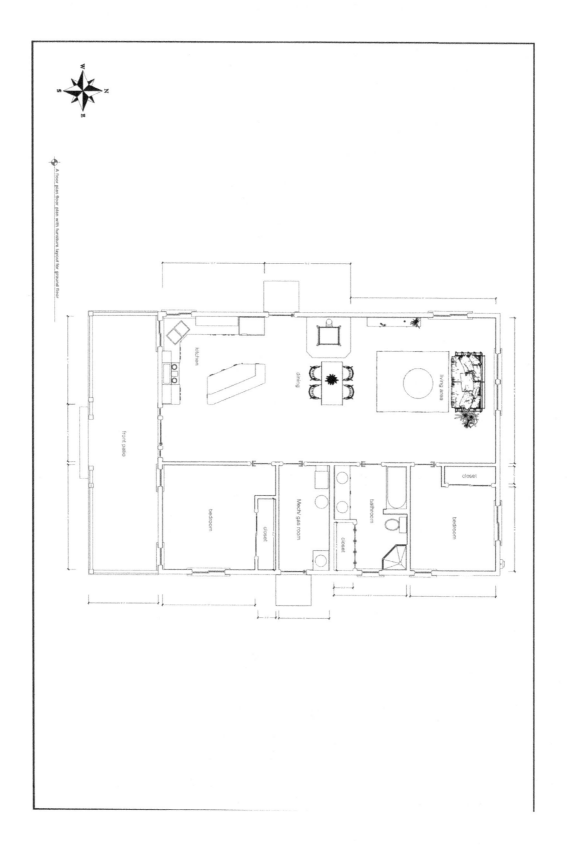

A floor plan floor plan with furniture layout for ground floor

kitchen

front patio

dining

living area

closet

bedroom

closet

Mech/ gas room

bathroom

closet

bedroom

74

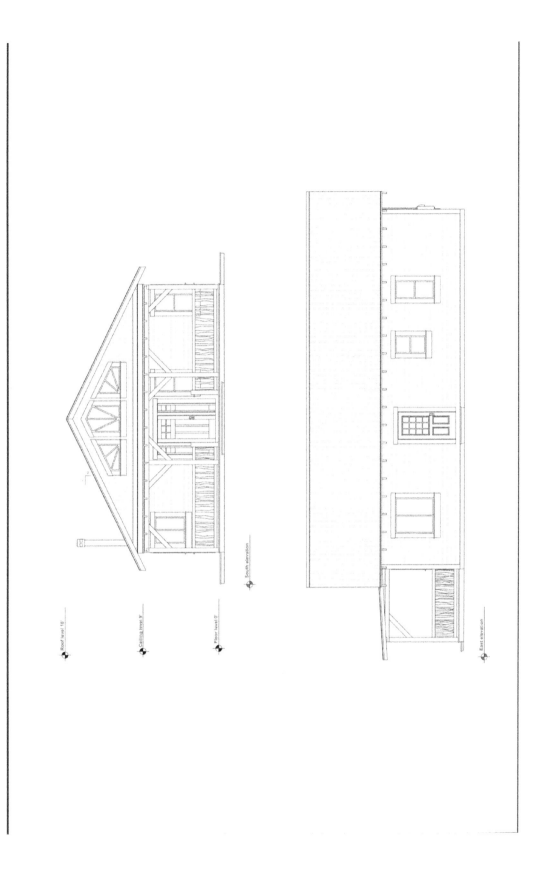

Roof level 18'

Ceiling level 9'

Floor level 0'

South elevation

East elevation

The Urban Nest

Welcome to "The Urban Retreat," a cutting-edge barndominium that effortlessly blends minimalist luxury with functional efficiency, all within a compact 930 square foot layout. This blueprint is perfectly suited for individuals or couples who champion minimalist living but are not willing to compromise on modern comforts and stylish aesthetics.

The design features an open plan that seamlessly integrates the living, dining, and kitchen areas, creating a sense of spaciousness that belies its modest footprint. The kitchen, equipped with state-of-the-art appliances, merges sleek design with practicality, making it a joy for cooking and entertaining alike. Adjacent to this central living space is the bedroom, a thoughtfully designed sanctuary that includes ample storage solutions and an integrated workspace, ideal for those embracing remote work or nurturing creative projects.

One of the standout features of The Urban Retreat is its expansive patio, accessible via large sliding doors from the living area. This outdoor extension of the home provides a versatile space for relaxation, al fresco dining, or hosting gatherings, effectively bringing the tranquility of nature right to your doorstep.

Moreover, this blueprint emphasizes sustainability and energy efficiency, incorporating materials and building techniques that reduce its environmental impact. The design not only meets the aesthetic and functional needs of modern homeowners but also aligns with the growing demand for eco-friendly living solutions.

The Urban Retreat is more than just a place to live—it's a lifestyle statement for those seeking a blend of urban efficiency and rural serenity, offering a sustainable, adaptable, and luxurious living environment. It represents an ideal solution for downsizing or simplifying life, emphasizing quality of life over square footage without sacrificing comfort or style.

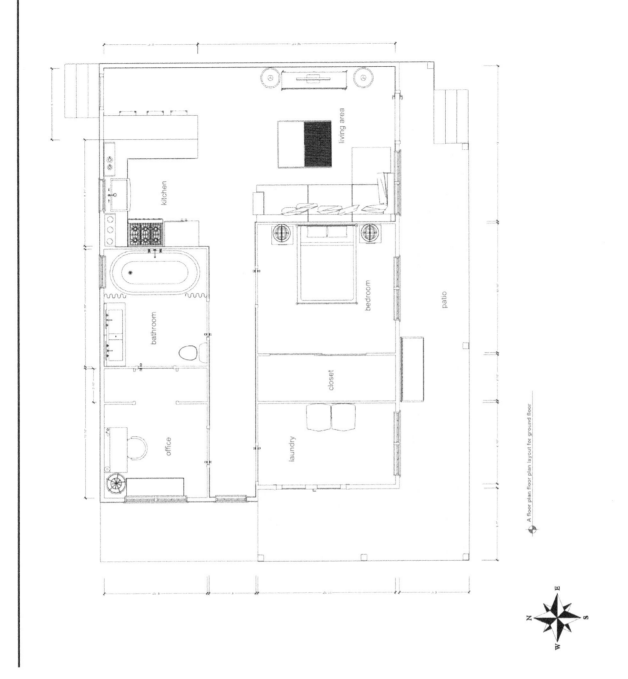

A floor plan floor plan layout for ground floor

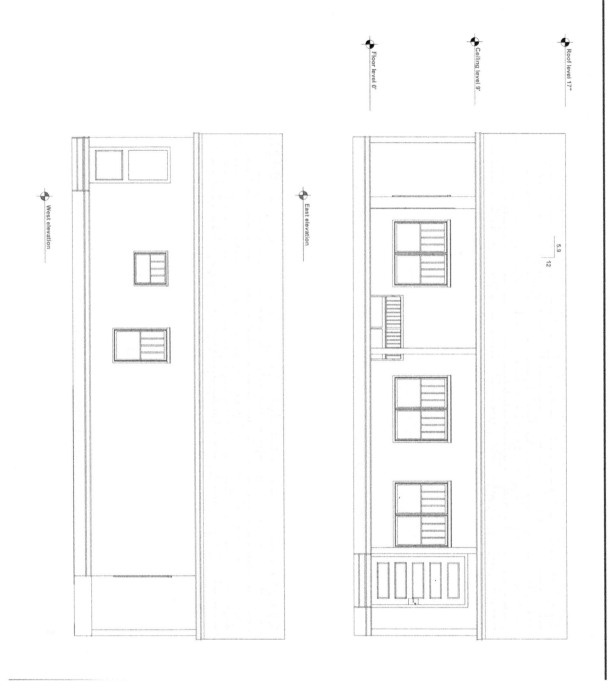

West elevation

East elevation

Roof level 17"

Ceiling level 9'

Floor level 0'

5.9
12

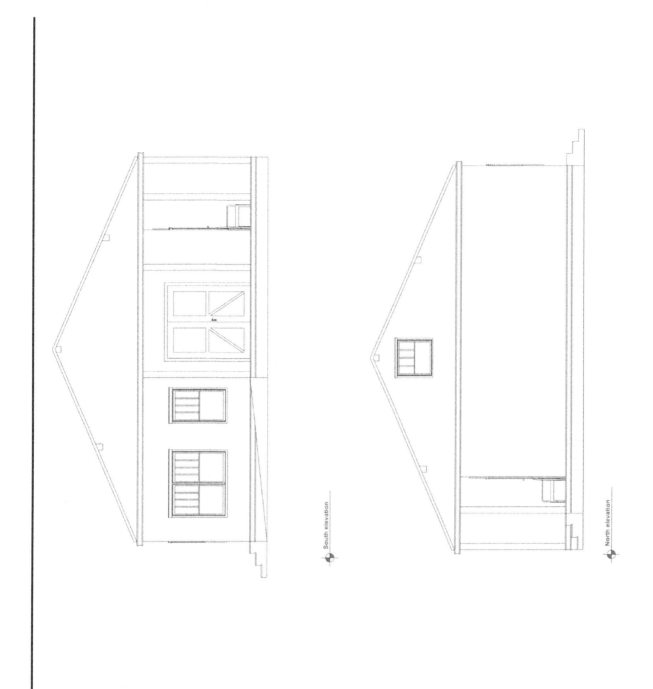

South elevation

North elevation

Conclusion

As you turn the final page of this guide, you're not just closing a book—you're stepping closer to realizing the dream of building your own barndominium. Throughout "Barndominium Building Mastery," we've traversed the comprehensive steps from ideation to completion, equipping you with the knowledge to bring your vision of rural elegance to life with confidence and creativity.

Building your barndominium is more than constructing a house; it's crafting a lifestyle that aligns with your aspirations for more space, freedom, and a connection to nature. Each chapter aimed to not only inform but also inspire, showing you how adaptable and innovative these structures can be. From the detailed blueprints to the practical budgeting strategies and the essential tips for maintenance, this guide was designed to ensure your journey is as fulfilling and trouble-free as possible.

Now, armed with a deep understanding of cost management, design customization, and construction techniques, you're ready to embark on this exciting venture. Remember, the unique floor plans and renderings included are just starting points. They are canvases awaiting your personal touch, ready to be transformed into homes that resonate with your personal style and functional needs. As you move forward, take with you the lessons on navigating regulatory paths, selecting the right materials for your climate, and managing your construction project efficiently. Let each choice reflect your commitment to quality and your desire for a home that is not only built to last but also built to change the way you live. Thank you for choosing this book as your companion on this transformative journey. May your new barndominium be a source of pride, a place of comfort, and a canvas for your dreams. Remember, the process is a significant part of the adventure. Embrace each step, anticipate challenges as opportunities for growth, and, above all, enjoy the journey of creating a home that truly reflects who you are.

Here's to building not just a structure, but a sanctuary—a place where memories will be made, where the walls will resonate with laughter, and where every corner will tell a story. Welcome home.

Scan the QR Code and access your 3 bonuses in digital format

Bonus 1: Basic Furnishing Guide for Your Barndominium

Bonus 2: Annual Maintenance Tips for Your Barndominium

Bonus 3: Utility Setup Guide for Your Barndominium

Made in the USA
Monee, IL
02 January 2025

75778915R10050